Danny Oceans de[illegible] profound and unique new vision upon spoken and written words. His work is all about the comprehension of communication and subsequently utilising the power of words by using intention-oriented speech patterns in daily expressions. He teaches the process of Words-Thoughts-Beliefs-Life, the mechanism of manifesting greater and more accurate results, by becoming aware of expressions and words and by adjusting and aligning them, in accordance with both the initial vision and nature's characteristics. Danny's intention is to reveal the various ways of improving any individual's life through the establishment of new habitual speech patterns that will be expressed as satisfying results in all aspects of life and are truly aligned with the initial intention of the speaker.

TO
LOVE
MAGIC
VANILLA

Danny Oceans

Talk Like Nature

AUSTIN MACAULEY PUBLISHERS™
LONDON • CAMBRIDGE • NEW YORK • SHARJAH

The story, experiences, and words are the author's alone.

A CIP catalogue record for this title is available from the British Library.

ISBN 9781035861361 (Paperback)
ISBN 9781035861378 (ePub e-book)

www.austinmacauley.com

First Published 2024
Austin Macauley Publishers Ltd®
1 Canada Square
Canary Wharf
London
E14 5AA

A supreme thanks to my beloved wife, Dr Viktoria Oceans, for supporting my philosophy, for editing this book, *Talk Like Nature*, and for believing wholeheartedly that the philosophy, the insights and the unique ideas and wisdom within these pages will - with high certainty - contribute to the reading individual and improve his/her life vitality.

Table of Contents

NATURE IS THE
SILENT GUIDE
OF HUMANITY
AND LIFE

Books by Danny Oceans

Life As We Speak

Words of Manifestation

Linguistic Potions

Yes, You Might Get

Talk Like Nature

Constructive Communication

Allowing Gratefulness

Visions

This Is God

Conscious Fabric

My Bless Book

Each word has the capacity to manifest a vision that leads towards an imagery within the mind. The individual gets life-supporting energies when imagery is fused with emotions and feelings; constructive words and visions are the source of beneficial life-supporting energies.

Many of the commonly used words are translated according to their unique preconception through life experiences and according to the ideas that the individual has adopted along his or her days with a solid agreement of them conveying truths about reality. In order to generate the most appropriate and desired imagery, the words that support the initial intentions in the most suitable manner become the very best choice in order to achieve the expected results.

The line between successful encounters of communication in life and not so successful ones is truly very fine, and most of the time one is very certain that his or her wording and construction of communication will surely lead towards what he or she has in mind as the outcome. Many words are so close to one another, that purely with solid awareness and true comprehension of their finer inner translations, will an individual be able to define the most constructive and suitable ones.

In this book, you introduce yourself to very significant pairs of words and are able to comprehend and reveal the unseen fine line.

'I become a successful communicator once I comprehend and utilise the offered word choices within *Talk Like Nature*.'

'My new way of utilising the harmonious ways of nature and the pure alignment with my language is the establishment of habitual aware communication and thinking patterns.'

Real Attributes

Life is about expansion, advancement, progress, continuity, manifestation and everlasting existence. Those are the real core characteristics of life and of my own existence. I am part of this manifestation like everyone else and all of those fundamental attributes and qualities are within me at every single moment, creating my life as I know it and establishing my world accordingly.

My language, and the use of it, is one effective way for me and the rest of my fellow men to manifest life as I know and experience. My words turn into thoughts, thought patterns and habitual mind focus patterns, which manifest what I sense as life.

The language is rich, beautiful and amazing with its variety. However, it is man-made. It came into existence with the process of conscious thinking and conscious awareness, and it is part of the objective world with the attempt to become a channel to the true core of my manifestation, the subjective realm.

All the words that I utilise on a daily basis manifest the life that I know and the life that I am aware of; thus leading me towards experiences, events, situations and circumstances that I would often describe as constructive and at other times as not so constructive at that given moment in time.

Words have special and built-in characteristics that are derived from the translation of my subconscious mind for every specific word. Whenever I use a specific word it is examined, monitored and evaluated within my subjective mind and will be associated with the memory of the life that I have lived so far.

This unique translation establishes the impact upon my overall picture of my personal existence in relation to life as a whole. When those words have more life and natural characteristics, the flow between my subjective realm and my outer environment becomes greater, stronger and in sync with the channels of the manifestation of life.

This establishes a very supportive stream of materialisation and thus life shows me much higher constructive events, circumstances, experiences and pictures.

Some of the words are in much greater harmony with the core characteristics of life and Nature, and therefore, manifest a much more harmonious life path for me to step upon and call: my life timeline.

Becoming selective with the words that I say grants me the benefit of living within a harmonious system. This refines the bridge between my conscious awareness, the objective world, and the subconscious awareness, the subjective world of my mind.

Once this bridge is in pure harmony, when it possesses a unity of the two parts that are in strong accord, my life becomes phenomenal, enjoyable, fulfilling and remarkable to the most apex capacities.

I am about to read words that are the preferred and better choices and the best life-creating elements for me to utilise as long as I can speak, read, listen and write.

Why am I doing this? This is a very easy question to answer: I do it, for I intend to become successful, fulfilled, healthy, peaceful, creative, abundant and wealthy.

Exactly as the life around me is the nature around all of what I can see, whenever my eyes are open, I become aligned with the force, which I am a part of, by using the absolute, most adequate, materialising words and speech patterns. My true awareness and ability to choose and concentrate upon a selected choice of expressions are a path towards the high materialisation of whatever I desire.

Help Support

I have probably used those two words many times in my life, looking at my friends, family members and many associates over the years, allowing them to know that I am here to guide and Support them.

On each and every occasion, my intention has been honourable, and my thoughts have been pure with the solid idea of improving their situation, thus elevating them towards more desirable grounds and heights.

Once the word Help is used, the person, whom my word is aimed at, begins to activate a process. This is an inner process, and it relates the word help to the necessity of assistance, to a situation where stability is not present; even to stages in the listener's life when success has not been visible.

Once the word support is used, the inner process will conclude that there is some kind of immediate and overall success within the individual's actions. It, of course, implies that there is some space for improvement; however, this is built upon constructive activities.

The word support highlights the successful actions, the worthiness of the individual and the appreciation of what has been done so far.

It boosts the very important self-confidence, self-worth and the valuable acknowledgement of the actions until that present moment. Those allow further actions; they maintain enthusiasm and excitement, and they allow me to keep on having faith in completing the assignment in a successful manner.

Once I offer someone support, I tell him or her that I admire that, which has been accomplished until now, that I appreciate its worth, that I am there to contribute from my capabilities in order to improve the appreciated structure or achievement and, most importantly, that I believe that it is possible to make it even better.

They hear, between the lines, that they are successful, appreciated and supported.

Support is definitely the better word to utilise. It represents that characteristic of life where everything that has been made and achieved so far is good and can even be improved by further progression.

It conveys the element of pure collaboration based upon partnership. The word help is constructive to a certain degree; yet, does not truly harbour the essential elements within it for inspiring the other party towards further action.

I prefer to express the word support and show the world around me that all that there is so far, is accepted, admired, appreciated and loved by me; I genuinely intend to utilise my talents in order to elevate All That Is unto even higher and more satisfying levels.

Nature is in a constant state of supporting its elements with growth, life, expression and manifestation, and it is not ever in a state of help, for it knows the true inner characteristic of its elements and cherishes within it the notion of them as strong elements that have the power and the capacity to lift themselves and to express themselves in a profound manner; all they can receive from Nature is kind Support.

Chance Opportunity

When I allow myself to express the word Chance, I suggest the following to my subconscious mind each and every single time I say it: 'It is out of my control and capacity', 'The situation is subjected to outer forces' and 'I am at the mercy of something outside of my powers and jurisdiction'.

In simple terms, it suggests that I will sit and wait to find out if the outcome is in my favour and as expected or if I should otherwise wear my unhappy face once the undesired results have arrived.

As I speak, my subconscious listens and follows my suggestions, both the conscious ones and the unconscious ones. It follows my orders faithfully. My preference of using the word chance in order to describe a situation that has a probability to unfold into a successful one is not the finest of guidelines.

It will not lead and direct my biggest power, my subconscious mind, towards the full materialisation of my vision, and will not guide it towards the precise actions that are in true alignment with my expected and desired outcomes.

Once my spoken and written words are those of Opportunity, the manifested inner picture begins to lead towards a unique road map for my subconscious mind to follow. It establishes a vision and a protocol to be followed, which will suggest the following: 'There is a great door of success right here, right now'.

'I am on the verge of materialising my visions', 'My action is pointing straight towards my desired outcome' and 'I was given a golden gift that I had better utilise in case I intend to experience greater success and fulfilment'. All of the above bring about a master plan of action for my truly faithful ally, my subconscious mind. It presents very clear guidelines of action for satisfying actions in all areas of life.

This establishes a great excitement and strengthens my hope and my belief in succeeding. This also works with others in the very same way once I speak this word or write it down for them. My best action here is to adopt the word opportunity and express it in those situations.

This provides me and others around me with a favourable mindset. It prepares all the visible and the apparently unseen resources and harnesses those for the completion of affluent results.

Chance is not part of life's characteristics. It is a man-made idea and, therefore, does not provide me with desirable results. It leads towards in-harmony and to places where all of my faith has been taken off its highest heights.

Life is giving me the opportunities to manifest, to expand, to express and to advance and *All That Is* is *What I Am*. Therefore, the word Opportunity is by far more related to the natural way of my existence and as such is my best preference.

Luck Success

I have heard many times that 'Luck has nothing to do with it', and I do tend to agree that Success has its own ground, foundation and mechanism, which is totally free of the attributes of Luck in the sense that most of us learn to comprehend it as.

Stating the word luck does of course deliver goodwill and does promote a nice connection and harmony between the two sides, me: the speaker and the listener. However, it is established in this manner for the reason that I, the speaker, would like to witness a constructive outcome for whom I communicate with.

Luck will be interpreted in a unique manner; yet, not at all close to my initial intention.

The subconscious mind gets messages such as 'It is out of my grasp', 'I could probably utilise an outer power to get where I would like to be', 'There is a big probability that I will not be getting what I envisioned' or 'I have no control and no power, and therefore, I have to be reliant on randomness'. This is definitely not the best message that I should get from my inner system prior to doing something new or important in life.

It will neither establish self-confidence nor boost faith nor hope. I will, instead of dwelling upon constructive and desired possibilities, focus upon the probabilities of not succeeding or having a satisfying outcome, which leads towards dwelling upon the dependency on outer elements and upon the gap between me and the process of life. Success, on the contrary, provides any individual with the very best tools inside the box in order to achieve worthwhile moments and events in life. Success is the exact word that paints the intended vision for the person at hand.

It has the capacity to establish the desirable focus upon the very best result of any action; to have me maintain the essential attention upon what I am about to accomplish and to boost my prominent self-confidence.

No matter what the upcoming actions or events are, success is a fine choice and leads straight towards trust, faith, constructive possibilities and outcomes exactly as was envisioned. Words are my tools to shape my mind and thoughts and lead towards the restructuring of the inner forces towards favourable results.

Success has the fine characteristics of life itself, for anything that I can possibly interact with is successful. All that which is around me has been manifested with pure intention and has become a favourable result for me to observe, reveal, enjoy and acknowledge.

Luck is the word that has no true relation to the way life and the overall existence of me operate and manifest, for luck does neither hold the blueprint of total expression nor the absolute totality of intention, which life and Nature have and operate by.

Nature manifests and very much acknowledges its own totality and its full ability to establish anything at any time, with solidity.

I state success for myself when I hear something happy, when I feel like doing something new when I set a clear focus upon fulfilling outcomes or anytime when I have a moment to elevate my faith, hope and self-confidence. I am the Lord of my mind and Success is a tremendous ally for me in order to achieve whatever my heart desires.

Change Manifest

Nature operates through expression and Manifestation. All the fascinating objects, sights, visions and features that I can see, feel, smell and taste are the pure result of constant progression towards everlasting expansion and expression.

Nature accepts all of its creations and transforms them through additional expression towards new structures, new manifestations and new and more relevant expressions. Change is not part of Nature, for the path that Nature takes is based purely and solely upon manifestation.

Change suggests that what is present, here and now, should be replaced, and that implies that it should be gone and most of the time not be recognised as something valid, valuable and constructive.

All is valid, valuable and constructive to a certain degree and contributes to the advancement, involvement and ongoing progression of life and all its participants. Manifestation is the sole action that transforms any moment, object or event into a new and advanced one. It is one of the core characteristics of life itself, and it is what I utilise within my objective and subjective life as well.

When I express the word change, I covertly suggest that what is now is in most cases neither any longer valid nor valuable nor constructive. By doing that, I remove the value of the object, the event, the person or the structure of which I speak.

Any time where I consider something as not valuable, not valid or not constructive, I feel it within my inner structure, and this reflects and restructures my own validity, value and constructiveness to a not as high extent or not as high levels as would otherwise be possible.

Whereas, when I express the word manifest as an alternative, I powerfully embrace the new vision whilst holding the validity, value and constructiveness of the subject at hand. I accept the great contribution towards my personal progress and my overall existence and, therefore, expand and manifest new structures, in the very same manner as Nature does. Manifest maintains the essential power of the current structure, and this grants strong foundations for the new experience and the upcoming vision.

Once my intentions are those of a strong new structure, solid new events and profound experiences, I better maintain that very admirable and appreciated validity, value and constructiveness of whatever that might be, and move forward together with those important essences, exactly like Nature. I solidify my new steps with strength, certainty and power, with true acceptance and admiration towards anything that I have engaged in throughout my life.

Manifestation is the pure truth that Nature knows and operates by. It rearranges all that is visible and has a part within the physical reality, and any reality for that matter. I am the product of manifestation and not of change, and so is everything that I come in contact with.

All movements are those of manifestation, and so am I. I am one with the absolute manifestation mode of Nature and of all existence. I operate in the Manifestation mode with all of my actions, thoughts, beliefs and observations.

Done Complete

'It is done'. I have been using this phrase more times than I can possibly count. However, as I realise the true meaning, the in-between-the-lines meaning, and the message that is streamed straight into my subjective world, I will not be using it anymore, as far as I am aware of that and as long as I prefer a more fulfilling life.

Nature is always in a stage of evolvement and constant manifestation. In an underlying comprehension, the word Done suggests that what has been there no longer has the means or the validity of advancement and expansion or does not require the necessity of any further action, which thus implies that it is not in a stage of manifestation. This is not the way Nature talks to us.

Nature is always on the verge of renewal and of manifestation, taking whatever is there and continuously streaming life into it through recreation and transformation. Once it is done, it suggests that it is not to be touched at all; it has no further necessity of examination or observation; it has its own realm and space of existence without anybody dealing with it any longer.

Complete suggests the idea of completion, of achievement and of successful manifestation. It is what Nature shows to me with every initiative, every action and every activity. Once completed, it still has a place for improvement and a thirst for further evolvement. It is how all things around me, as well as myself, perpetuate their existence.

We are constantly with a thirst for further growth, expansion and improvement. Even while I feel complete, I always allow a stream of innovation and continuity to be part of what I am, and therefore, I am complete at every single moment and never done.

Whilst I maintain the focus upon completion, I allow new lines to be built, lines that will take the authority of continuing the journey of manifestation and expansion. Complete allows further action and grants new seeds to evolve and to grow. It is like having new employees and envisioning their future in higher management as the captains and the navigators of the upcoming years and actions.

When I look around me in the neighbourhood where I lead my life, it is always completed to a certain degree, yet never done. All the plants, the people, the businesses, the organisations and the systems are in a state of continuous improvement, development, progression and natural growth.

I am the same and so should my usage of language be. Aligning my language with the pure nature of Nature is for my biggest benefit and keeps my overall structure as harmonious as possible.

Each word either generates harmony or removes harmony. Nature is all about generating harmony, and this is how I can measure what my very best preference of words is. I feel the effect of each and every word explicitly upon me and my surroundings, and in a situation where the conclusion is that of more harmony, the approval of that word will have been decided.

I am authorised to maintain a harmonious environment for me to grow, evolve and improve in by knowing that I am Complete, and further energy will be available for my personal expansion towards improvement.

Enough Sufficient

The right application of thought is my way towards success, health, wealth and true value fulfilment. My thoughts are the foundation of my experiences; my words are the raw material for building those powerful thoughts.

Enough represents situations where there is no necessity for more, and anything more, than what has been given, will do no good for the current situation. It is a word that I used to use when my mood was not very calm or when peace was not present.

Most of the time I find myself peaceful and calm and yet utilising the word enough and thereby establishing the completion of the flow that may have continued otherwise. Life is a constant flow towards anything and anyone. It sends a flow of riches, energy and elements for creating realities, events and life experiences through a continuous process.

To tell life enough is to suggest that I am no longer in a state where I can utilise life's material and supplies, and this removes me from the progressive stream of personal expression, away from plenty.

Flow is what I am, and all my organs and my systems continue their actions until the concluding physical breath. Once the word Sufficient is utilised, there is an acknowledgement of acceptance of what has been provided so far and an acknowledgement of gratitude for what has been given and received.

The source of the supplies will be granted the required appreciation and the acknowledgement of its validity and value for the latest actions. This streams sufficient energy towards maintaining the upcoming supplies and actions from that party.

By using the word sufficient, I signal the highest manifestation of core thoughts of appreciation, and this supports me very well once there is a place for getting more of what has been given.

The everlasting state of life and the true nature of all systems around us, such as the solar path around Earth, the continuous movement of water from Earth to sky and back to Earth again and the breath that I keep on taking are all the visible characteristics of all that I am, and all that there is.

My thoughts and words, once aligned with those natural traits, do have a finer place in my existence and are a better match for all of my actions and activities. Nature talks, and I absorb it with my eyes, my ears, my hands, my nose and my heart.

Once Sufficient is the utilised word, all of my surroundings are able to permeate further into my inner realms and be more than happy to continue supporting me, for I establish further support and flow of supply by using this word, for it is talking exactly like Nature.

Closed Signed

When a new deal or agreement reaches the state where two parties accept the written conditions, a new flow of actions and of activities is manifested. Both sides participate in one of the most significant attributes of Nature: collaboration.

When this is reached, the word Closed represents the idea of 'The deal is closed', which will not reflect the actual value of this new materialisation.

It will not even remotely support the state of evolvement and of expansion. This word will purely state a clear focus upon a state of mind and thoughts where no open road is to be found, no clear and stable path is in sight, and no way is visible.

Nature continues its movement, no matter what the situation is.

It continues to flow and move like water does, finding and reaching the next available path. The available energy had better be maintained, especially once collaboration has been proclaimed for the benefit and success of all sides.

When I conclude a new collaboration with the use of the word Signed, I maintain an open path towards further actions; thus, keeping a clear road and establishing the idea and thought of a new beginning, which does require and greatly benefits from an open course.

Signed is preferred for the benefit of the new establishment as it stems from the root sign, a way where Nature shows me my pure nature of actions and existence.

Once the idea of signed and signs is the source of my thoughts and my dwellings, all doors are available, confidence is in sight and faith becomes the leading force supporting my new vision.

Keeping all my activities open by no longer using the word closed once a deal is established, will have my actions result in an open path and will direct the flow of ideas and improvements towards the successful outcome of what has been agreed upon by both parties, in this case the deal signed by me and my new partners.

I use the phrases: 'It has been signed', 'We have reached a solid agreement', 'Everything is Signed and ready to be exercised', 'We are open for business' or 'Another successful deal'. With every door that I go through, I keep on opening additional doors, similar to Nature.

Finish Conclude

The elements of everlasting action towards advancement, expansion and pure expression are the traits of Nature and the guidelines of my existence. I am here to express my talents and abilities; and, by doing that, grow, expand and advance.

Once the word Finish is used, I suggest that 'There is no further way of growth', 'There are no new ideas of expansion' and 'There is no desire for further advancement'.

This is not close to any of my true Nature-given attributes and is not aligned with Nature itself, which I am a valid part of. The word finish suggests that all actions must be halted and that all the available power of manifestation should be set aside for a while. With this, Nature will neither agree nor will it support this, and neither do I.

Once my previous thoughts and actions are Concluded, a new horizon is established and realised, manifesting the place where all the agreed conclusions are pointing towards and where all can be realised.

This opens a new realm for practising what has been established in the blueprint, which harbours true similarities to Nature. When I have completed my actions and all of my assignments, I have concluded this stage and am looking forward to the next one and towards new assignments.

When using the word finish, the suggestions point to no further growth and to no further desire for action.

Conclude is by far the better option for me in order to keep on being active, being valuable and being a contributing part of the process of life itself, as well as my own existence.

My subjective mind receives streams of ideas about value, advancement and opportunities when it deciphers the term conclude.

It gathers all the related experiences, ideas, thoughts and ways towards completion whilst maintaining its focus upon Concluding.

Anything that can ignite further action, excitement and preparation for new activities becomes aligned with life's and Nature's characteristics. Those will be the favoured and the most suitable ones if the intention is to acquire success, health, wealth and advancement.

Many Plenty

All that I experience in life is the result of what I believe to be presently true and to have been manifested for me by the immediate environment.

I have the tendency, as a conscious being in this physical framework, to quantify things and to count them as if everything will get greater value with an assigned higher number. I do that on a constant basis and count and quantify all that I observe whilst I am awake and aware.

The quantification of things is mainly based and built upon the numerical source that I have been taught since birth. The numeric quantification allows me, so I believe, to know and comprehend the greatness of the matter.

A higher number of apples is, in most cases, better than not-as-high a number of apples; it, of course, depends upon my current situation. In the case that I would like to bake a large apple pie for guests, a load of apples would be the preferred condition.

I do count and once I count, I assume. The assumption that I give to the counted number is based upon what I have been taught to believe to be a lot or not so much.

Once the quantification is in place my conscious mind begins to balance the quantity according to what I believe to be the right amount or number and according to my definition of a lot or not so much. In a number of these situations, I tend to utilise the word Many as the one to describe a large number of something.

The word many describes a large number; however, can suggest that it might be too much of something as well, and therefore, allows me to initiate the process of halting the production and the progress of what I am engaged in.

It is consciously determined what is too much and what would be the right amount of anything. I quantify and utilise the word many in order to manage the process of manifestation.

Nature does not ever manage the process of manifestation, for it has another idea of what should be, how many of it there should be and what would be its position. Nature allows itself to channel the eternal amount of creating energy in order to maintain a profound presence and life.

Nature operates through the guises of eternity and through a non-numerical pattern. It allows the blossoming of all of its parts in an instinctive manner, it allows Plenty.

Nature radiates plenty. Nature shows plenty. Nature is plenty and not many. Plenty is the innocent acceptance of continuous manifestation with the notion that all is accepted and that the core of the creative energy is eternal, and therefore, has no beginning or ending.

Plenty is the most pure manner of observing Nature which allows the observer to see how Nature sets quantification aside and allows plenty and energetic and natural growth to take place.

Whilst I teach my mind these elements of Nature where I set aside the numerical system, I allow pure manifestation to take place whilst allowing any number of things to be present and to be manifested; the balance in my life will grow bigger and the harmonious interaction with things will naturally take place and reveal itself to me in this realm of expression. I can benefit from setting measurements, such as time and quantity, aside.

I will run better, faster and with greater endurance and agility when I do not pay conscious attention to the distance or the time. I will manifest more and with greater efficiency when there are no defined quantifiable elements to work by. I am free whilst I see, observe and grasp all that I have, do, manifest, receive and experience as Plenty.

Spend Utilise

The true nature of Nature and life is being the active and physical materialised part of the Prime Source, and there is solely a pure and constructive Utilisation of what is available. The Prime Source is the core that all existences, realities and manifestations emerge from, and I, as well as Nature, am its physical agent; living, creating and experiencing a life within physical dimension.

There is solely one proper way of using all the profound ingredients that compose life as I know it, which reconstruct it on an ongoing basis and that form into actual objects. Nothing can be used in a non-constructive way and all remains to be forever available in regard to getting a new form or a new shape whilst creating a new scene for all the observers who are around it.

The utilisation of the word Spend indicates a state where a non-constructive use of the item at hand has taken place. It tells my subjective mind that the action has not been appropriate, not constructive and has not been utilised or exercised as it should have been. It also states that what has been is no longer available, no longer accessible to me and/or cannot be restored.

All these implications of the inner meaning of the word spend have no traits of the true nature of my reality and are neither aligned with the way Nature talks to me nor the way it operates.

I do have access to anything and everything at any time. I can achieve constructive outcomes from all the experiences I have had in life, and I can build a solid new platform of advancement. The Conscious Units, which are the active unseen particles available all around me, are constantly transformed into new shapes. It is the true nature of things.

Everything is transformed again and many more times according to the clear instructions of its unique governing mind, which guides it into materialisation and into an objective structure. As I observe Nature, everything is part of a cycle and gets new forms, establishing continuous and everlasting existences.

The trees lay their leaves, which in turn become soil, which thus nourishes the tree via the roots and the trunk in order for the tree to grow its leaves again and nurture other vegetation.

All is utilised; nothing can be spent. Utilisation is the characteristic of life and as such it is in my best interest to express the term utilise with the inner comprehension and the valid relation to what I am and to all that is around me.

Once I comprehend that nothing can be spent, and once I implement the idea of sole utilisation within my system of thoughts and beliefs, my way of relating to events is saturated greatly with acceptance, approval and unity. All is here and will be around for eternity. The material and the fabric of all that I experience have formed many shapes, many figures and many sights and continue to do so in the everlasting cycle of life.

Allowing the word Utilise to be the guide within my speaking patterns, directs my mind into this insightful revelation. It shapes my thinking patterns about how life works for me and through me. My thoughts are the essence of my experiences. My selective words and speaking patterns are the base for having the most intended, most constructive and best-performing thought patterns.

Serve Deliver

Life Delivers. Life is the master and remains the ever-leading force of All That Is. Life masters, materialises, manifests and manifests everything, and therefore, it delivers results to anything that it involves within its realm of actions.

Serving is usually immediately interpreted as a state in which one party is ruling over another, which automatically implies that the other party does not have an equal value compared to the former.

Life is available to us all, and I have the very same powers and forces as everybody else; powers to structure my reality as I wish; forces to manifest whatever I feel is right for me at any given moment and the absolute capacity to manifest what I have a desire for, what I would like to do and what I would like to become.

I am life and so is everybody else. I am here to deliver and not to serve, for I am equal to all that I can fathom with my senses, and all who grasp me are equal and have the very same attributes and traits for performing the very same magical actions as me.

Serve is not my choice, for I like to feed my mind with equality, with actions by will and activities by choice. Deliberate alignment with the traits of Nature is my free volition to shape life as I desire, to build as I believe and to manifest as I dream and envision.

My responsibility is to maintain a clear mind, a strong realm of thought and a contributing structure, which maintains the right relation to life, to Nature and to myself.

I am in the midst of all actions, and I deliver and stay fruitful towards what is around me with equal strength and status. Nature is the force that delivers life, manifestation, expansion and advancement, and so I am here to get closer to that which supports and delivers existence to me. Life delivers to me all that I require for continuing my progress towards true comprehension and greater growth.

All is given and handed with full intention, with enjoyment, with clarity and with excitement. All, that may be even better for the purpose of my growth, is ready for me, here and now, and with supreme thrill by the delivering force. All that I can do in order to live by those attributes and characteristics is to become selective of the seeds that I choose and plant and by which my mental garden is cultivated.

Comprehending the true meaning of deliver is the proper seed for this sublime rule. 'You know how to deliver' could be a sentence to be utilised when someone has provided or delivered satisfying actions and results for or to me.

'I love the delivery here and appreciate its efficiency' leads me towards an overall refinement of my mental garden through the use of the finest words and terms. I am here to deliver and never to serve. I am here to manifest and utilise my sublime talents and allow those to be greater and stronger, and to solidify my virtues whilst experiencing life like all the people and Nature around me.

All is here to Deliver from its wisdom, capabilities and talents and to express itself with acceptance and with its own true volition. Granting myself this notion gets streamed into the minds of all whom I contact and engage with; thus, rewarding and gratifying them with the very similar precious insight.

Manipulate Manage

Manipulation usually has more of an unconstructive connotation and hints to the direction where I control and force the other to do something as wished and this, most of the time, unwillingly.

I already get unfavourable vibrations simply by saying this word, for it feeds my mind and my awareness with this ungodly action. The translation of it is what I have been taught to comprehend, and it will most often be grasped similarly by the ears of most people to whom I state it.

My intention, with any communication, is to allow the support of the other and have him or her be on board with whatever I would like to achieve. Whenever the word manipulate is utilised, disharmonious vibrations are present and those will not lift the current level of quality.

My very best option, with high certainty, is to utilise the word Manage instead when addressing one person or more, either when talking or when writing to them. This way trust is built and benefits me, for it will enhance and support any intended collaboration.

This form of communication and those channels of direct communication via the spoken or handwritten word, which I prefer and choose, have a strong effect, and therefore, the related outcomes will be visible soon, for the term manage generates the appropriate associations in the mind after the ears and eyes have deciphered its meaning.

It will immediately be related to other terms such as support, guidance, collaboration, intuition and manifestation; both consciously and subconsciously. These ideas are very subjective from one person to another; however, they are close relatives of manage, and thus, convey similar characteristics and traits.

Here, I have presented what I relate the use of the word manage to, and I am very certain that most of those terms will jump into the minds of many others, and therefore, establish constructive body and mind reactions, vibrations and thoughts.

To succeed with my intention and get others to support me and be willing to join my ideas, projects and suggestions, manage is, positively, always the finest choice.

Using statements such as 'I will be able to manage that properly', 'It is manageable' or 'I managed to...' amplify and elevate my capabilities to succeed and expand my sense of self-worthiness, my self-esteem and my faith in successful actions and activities.

Nature manages all existences and operates within the process of continuous guidance, all towards higher intelligence, advancement and capabilities.

The word Manage contains and conveys those ideas, characteristics and traits and streams them all towards me whenever I set a clear focus upon it and utilise it. It is the preferred word for getting a pure and solid collaboration and performance.

Wish Have

Wishing to have something makes me remain in the situation where this thing, that I wish to materialise, maintains its momentary position, which is away from me.

Wish signals the mind to structure a picture of me on the one hand and the precious desired thing on the other, and this will have my mind dwell upon the space between the two, thus recreating it (the space) for future events and circumstances.

Each word counts and the word wish will always support the current situation, gathering the associated envisions, setting the focus upon not having what I would like to Have, do or become.

I wish for nothing; I have it here and now within my mind. Have supports my purpose better, for this sublime word represents a moment where all is already achieved and materialised. It manifests a vision in anyone's mind of getting to the target, materialising the golden visions and getting all desires fulfilled.

I tell myself a few times that 'I wish I had more wisdom' and then feel the body's reaction to the stated statement. Later on, I will tell myself that 'I have more wisdom' a few times and measure the appearing reaction and vibrations. The latter suggests that I do have a certain level of wisdom. Therefore, I get the feeling of my current wisdom much easier and get attached to it instantaneously.

The idea of having more wisdom, and the movement and the flow of wisdom into my being is conveyed by this statement, allowing me to feel the reality of having more wisdom in the present moment.

This maintains the focus upon my thoughts and ideas about my wisdom, thus increasing the amount of wisdom progressively. The first statement ('I wish I had more wisdom') triggers a sub-reaction within my inner self, allowing me to know that 'what I have, has no real value'.

It will even remove some of its value due to the derived focus and association with the sentence itself and its core word, wish. Wish does not take me anywhere where I would like to be. It keeps me at the current level and often takes a part of whatever I already have and evaporates significant portions of it.

Nature speaks with confidence, trust, abundance and plenty. The word have represents those attributes and signals them directly to my subjective mind, thus creating a situation exactly as I prefer.

Getting in sync with the way Nature talks to me and behaves is my new elevation towards a fulfilling life. Acting like Nature through linguistic habitual patterns is my finest choice, which, in this example, is to state what I would like to Have.

I replace statements such as 'I wish you were here' with 'I would love to have you here' and 'I wish I was taller' with 'I would very much like to have that height'. The word Have is very creative, an instantaneous action towards manifestation. It triggers actualisation and is the ignite of personal expression; so, I allow myself to utilise it as often as I can.

Basic Fundamental

I like to be very pleased with what I suggest to others, and often, I am very happy with what I have stated and pointed out as the right and most appropriate direction to and for the people around me.

The truth is that each word of mine contains the seed for a bigger and more vivid structure, which influences and shapes the way others act and continue their current actions.

The word Basic is translated as something that is closer to the mid-levels of the scale. It represents the middle level, the in-between or even the levels closer to the absolute baseline. It is neither aligned with progression nor with advancement.

It does not manifest success and innovation; it directs the mental structure to set a clear focus upon the baseline, namely the bottom. The word Fundamental on the other hand triggers ideas of necessity, of value, as well as of worthiness within my subconscious.

This empowers and encourages the people whom I talk to, to think thoughts of necessity, value and worthiness, for their mental structure derives those qualities from the word that they have literally deciphered with their ears and minds a moment before.

Nature uses fundamental elements; that is the way it operates. It uses key ingredients in order to perform expansion, advancement and manifestation. Nature is continuously talking to me in a manner of value, of necessity and of worth.

Fundamental ideas convey those properties and contribute greatly to the minds of the people who intensely dwell upon them, and thus, align their line of thoughts with those of Nature. Once I consciously think of fundamental, I sense importance, vitality and essentiality.

My subconscious mind ties the derived connotations of the essence of the word fundamental to all the events and experiences within my life; thus representing the related terms in my life with the vividness of physical reality, where my body currently operates and resides.

Fundamental is the preferred and more value-providing choice. It contains the core elements of Nature and its way of behaving. It allows my mind to think in a way that has much in common with Nature around me, and thus, allows the thinker to compose a more powerful connection between the thought, the derived action and the participating forces of manifestation.

My words are the key for creating and maintaining the continuous connection with the architect of all realities.

Fix Improve

When I am about to Fix something, I suggest that there is something not very well with the current thing. There must be a very unpleasing situation that had better be addressed with seriousness, emergency and certainty.

By utilising this expression, I send a message, to those who are involved, that shows an undermining and downgrading of the worthiness of the object at hand, the situation and the people themselves.

This, however, does not allow any of the participants to reach more successful planes and levels.

It would rather promote an even greater gap within the entire group, and the individuals, who are involved, will dwell upon this apparent inefficiency of theirs, which has been mentioned and projected within the subconscious mind.

Using the word Improve as an alternative solidifies the trust, which I prefer, in the result or the situation and gathers confidence through the recognition of the effectiveness and the already brought good result.

Improve suggests trust in and acceptance of what has been manifested so far and is aligned with Nature's attributes and methods of manifestation.

I am able to have strong harmonious feelings solely by knowing that what has been achieved so far has value, and thus, I am as well valuable and appreciated, granted by my previous actions. My better choice, in order to promote a strong foundation for further growth and advancement, is definitely the favour of the words improve and improvement.

These support greater levels about to be reached, and the pure acceptance and the acknowledgement of value for all the people involved.

'It is time for improvement' is a very sufficient and profound statement, once. I would like to experience something new and greater, based upon the current constructiveness. Nature manifests and builds upon acceptance and approval, and I can do so as well by the constant utilisation of the word Improve.

Any time that I truly use this word with my profound awareness of the entire process, which is actively present inside my subjective and objective world, I set a stronger starting point with every one of my actions. I solidify a trustworthy foundation to build upon whatever I envision in my mind.

Powerful Statements

- Some of the words are in much greater harmony with the core characteristics of life and Nature, and therefore, manifest a much more harmonious life path for me to step upon and call: my life timeline.
- The word Support highlights the successful actions, the worthiness of the individual and the appreciation of what has been done so far.
- Life is giving me opportunities to manifest, to expand, to express and to advance, and *All That Is* is *What I Am*.
- Success is the correct word to be used in order to generate the intended vision of and for the person at hand.
- Success has the fine characteristics of life itself, for anything that I can possibly interact with is successful.
- I am the Lord of my mind and Success is a tremendous ally for me in order to achieve whatever my heart desires.
- Manifestation is the sole action that transforms any moment, object or event into a new and advanced one.

- Manifest maintains the essential power of the current structure, and this grants strong foundations for the new experience and the upcoming vision.
- I am the product of Manifestation and not of change, and so is everything that I come in contact with.
- Nature is always on the verge of renewal and of Manifestation, taking whatever is there and continuously streaming life into it through recreation and transformation.
- Complete suggests the idea of completion, of achievement and of successful manifestation. It is what Nature shows to me with every initiative, every action and every activity.
- Complete allows further action and grants new seeds to evolve and to grow.
- Aligning my language with the pure nature of Nature is for my biggest benefit and keeps my overall structure as harmonious as possible.
- Once Sufficient is the utilised word, all of my surroundings are able to permeate further into my inner realms and be more than happy to continue supporting me.
- Signed stems from the root sign, a way where Nature shows me my pure nature of actions and existence.
- The elements of everlasting action towards advancement, expansion and pure expression are the traits of Nature and the guidelines of my existence.

- My subjective mind receives streams of ideas about value, advancement and opportunities when it deciphers the term Conclude.
- The Conscious Units, which are the active unseen particles available all around me, are constantly transformed into new shapes.
- My selective words and speaking patterns are the base for having the most intended, most constructive and best-performing thought pattern.
- Deliberate alignment with the traits of Nature is my free volition to shape life as I desire, to build as I believe and to manifest as I dream and envision.
- All is here to Deliver from its wisdom, capabilities and talents and to express itself with acceptance and with its own true volition.
- Nature manages all existences and operates within the process of continuous guidance, all towards higher intelligence, advancement and capabilities.
- Nature speaks with confidence, trust, abundance and Plenty.

Let Allow

When I am about to do anything, as I believe to be constructive to my life according to my current progress and state, my thoughts guide me towards making decisions and creating plans towards the realisation of what I have in mind.

The freedom of action would be a very fundamental element in case I would like to experience that vision completely, vividly and fast. Each and every word counts, and therefore, I favour the most harmonious words that signal the element of freedom-of-manifestation towards my subjective mind.

Stating: 'I'll let it be' is translated into 'I will be kind enough to grant the freedom-to-be to whatever or whoever it is'; however, it also means that 'I have full independence and the power of decision-making on this matter'.

In the case of using the word Let, I have full governing power over who is going to do what, and thus, the true connection to things and to the oneness of manifestation no longer resides within my speech.

Nature Allows me to grow freely, to explore, to manifest, to become whatever I would like to become and do, without any supervision, and with complete freedom of action. Nature allows me to be whatever I would like to become. Nature allows the plants to be as they are and allows them to develop any aspect of themselves into new stages.

Allowing is the true fundamental nature of Nature, and I become more creative whilst living in accordance with the word allow. I become more successful and happier when utilising this very unique and significant attribute in my life with the assistance of my habitual speech patterns.

Stating 'I allow it to become' is saturated with the elements of freedom and of acceptance. It does mean that whatever is has full freedom of expression and does neither require any approval nor permission from anyone, myself included.

Allow means, that I allow it to myself. Once I am in the thinking state of complete freedom-of-expression, whilst radiating it towards all that exists, I manifest that state within my own core of existence.

My personal self-image, which is dynamically harboured within my subjective mind, evolves through those traits of life. This always grants supreme ideas and thoughts that are completely aligned with the true meaning of the word allow.

I become freer once allow is strongly implemented within my system, the system that all life experiences and events emerge from. My way of speaking towards all that apparently resides outside of my physical body, is the way I teach my inner self to believe what life is all about and that type of environment I prefer to realise.

My personal image evolves on an ongoing basis with each and every thought and idea that is transmitted to the subjective realm. My choices of words are the foundations of this precious image. I allow the word allow to present new traits and to improve my channels of manifestation.

The associated ideas related to the word allow support the new building blocks of the inner image; channelling new beliefs about how and what my life is from the moment I use it.

Allow is the supportive word and idea of complete freedom that is here in the first place and is not granted by anything or anyone else. I was born, and I have manifested and still do with the complete freedom-of-expression and freedom-of-manifestation, the same as all things in Nature. It is Allowed and always will be.

Deserve Worthy

When I feel the passion for everything, I establish the essential position in order to extract and exercise what I love the most and have a big attraction for, which establishes a state of strength and hope in my mind.

All of my actions are performed much better and very close to the fullest of their capacities once I have the feeling of Worthiness and the sense of value within my mind.

When I hear the word Deserve, I get a slight feeling of having to justify my current state. In my mind, this elaborates on the necessity of having to earn the right to do something before actually having this privilege.

Being and feeling worthy is what Nature intended to be a solid part of my identity. Being worthy is having the right to do anything once the idea is within my mind and having the right to have whatever is on my mind without the requirement of any prior actions or qualifications.

I am qualified here and now; I am worthy of my present, my life, my existence and my thoughts and ideas.

Nature looks at us as an innate and interwoven part of itself. I am worthy at this present moment, and all of my actions are already approved once the blueprint-of-thought is established and held.

Worthiness is the key word of accepting anything and anyone as an entity with the full right of expression and fulfilment. I am Nature, and I am worthy of whatever I see as my true and desired expression. When I look all around me into the vast landscapes of this spectacular planet, observing and experiencing the various types of scenery, sights and structures, I see worthiness; I see complete acceptance of true expression and vitality.

Utilising statements such as 'I trust myself to be worthy of that', 'I know that you are worthy of your desires' or 'I am worthy of all that I have and allow myself to receive' introduces a sense of high self-acceptance and trust in Nature.

Removing the word deserve and replacing it permanently with worthy, maintains the true collaborative relation between myself and others. They will all see me as a profound partner, and this will support great results. Worthy is a clear equivalent to the operation of the creative mode that is the nature of Nature.

All I see possesses the right of being expressed freely and totally without having to prove the right of doing so, and therefore, being Worthy of something is very aligned with the way Nature talks and structures my existence with everything I experience along my days.

Buy Afford

I can Afford what is available to me within this spectrum of existence. My vision outward is the evidence that I have the capability to have, possess and enjoy anything that is reflected and visible in the realm of the physical environment.

The objective world is the reflection and the dream of the inner subjective world, and my rights within this created world are to enjoy them to the fullest. I can afford and do not have to Buy something as if it is being taken away from someone else for a specific return and exchange of value.

All that Nature offers me is a gift; free to be obtained; free to be embraced. The animals and the trees do not pay for their place of living. When I think of the word buy, I assume a prerequisite obligation of earning in order to get what I would like to acquire.

Once Nature expresses its ideas towards its descendants, it holds the idea of rightfulness and worthiness within its consciousness.

There is one necessity to my current self and that is to express my uniqueness to the fullest of my capacities. I can afford' and have been granted the right to be able to afford whatever I desire and whatever has the urge of being expressed through my physical existence.

All is here for me to enjoy and to use as was emerged by the Prime Creator. Nature addresses the meaning of the word afford through its expression, and this is the most harmonious way of allowing myself to acknowledge that my being here is appreciated and valued.

My speaking patterns should better contain the word afford when describing the action of getting something new into my possession. The ideas that are sent into my subjective world empower the fact that all can be materialised, all is accessible, and all is appreciated by Nature and by the Creator in relation to all that I can see, taste, smell, touch and hear.

The ability to afford maintains a position of power and competency, surpassing the level of the necessity of having to pay for what is desired.

The level of Affording is the true alignment of my ideas and mind-structure with the fashion of Nature's expressions and talk.

Task Contribution

All the things that I observe around me have the very same core principles; they are expressions of love and excitement. All is here for the greater good and for the apparent future advancement of all living life.

Our actions are with the intention of pure Contribution. I am here on Earth and what I do contributes to and supports life and all parts of Nature. Once I describe any of my actions as a Task, I harbour the element of obligation, the essence of a must and of responsibility within my very own statements.

Nature loves all of its actions and does everything with pure joy, great vision, pleasure and complete acceptance. Nature does neither have tasks nor any nuance of obligation towards its progress, movements and events. All is manifested with love, excitement, true intention and curiosity towards the upcoming outcome.

Anything that we do with the very same characteristics of contribution is delivered with far greater results and electromagnetic vibrations; aligned with Nature's behaviour and strength'.

Contribution is one for all whilst knowing that one *is* all. Having the descriptions of my actions aligned with the element of contribution aligns me with Nature, empowers my daily activities and sets my mind up towards the truth that 'all is one' so that each and every one of my actions is a direct and a profound contribution.

I will now think of the word task for a moment and, afterwards, I will think of the word Contribution. I will now feel the vibrations within my physical body and the energy that flows through all of my organs whilst dwelling upon those two words individually.

I have noticed for myself that one of these two feels better and that one of the two is more harmonious and more life-oriented. In the case that I have a significant observation and can comprehend through my immediate feelings, which one is more constructive more life-oriented and Nature-oriented, I have found my answer. Utilising it often grants health, power, vitality and meaning.

In this way, I can obviously, and with ease, measure all the words and terms that I use during my day. Each word has a vibration, and it could be constructive and life-supportive to a certain degree or not.

The natural me benefits greatly from using and implementing the highly constructive and more harmonious words and terms. I can make a conscious decision to allocate more vitality and harmony by using apex words.

Outstanding Excellent

The most interesting unseen concept of the word Outstanding is that the subject at hand is described as something that is actually standing out. Many would state that it is a constructive element and ask why I would not consider it being a characteristic word of Nature and a related word to it.

The word out is, in many cases, grasped in relation to something that has been the pure result of a non-constructive activity, and in relation to the exact opposite of reward. When a kid in class does not behave well, it will be sent out with neither a smile nor happy vibrations by the teacher.

At ball games, the phrase 'You are out' is the result of an insufficient performance of the player, and therefore, he or she is no longer allowed to continue or participate further in the game.

My conscious mind grasps the term outstanding as a constructive one; however, as it goes into my Factory of Manifestation, the subconscious mind, it is divided into two words by a process of close cxamination; and then comprehended as out and standing.

In most cases, this allocates, as described before, unpleasant connotations and is aligned with inharmonious vibrations, and thus, contributes no further to any better outcome.

Excellent is pure. It is totally clear from any such unseen investigations, and my subconscious mind, the Factory of Manifestation as noted before, relates constructive events and ideas to the word excellent and relates it to further progress as well as to top and apex performance.

The word excellent harbours the word excel within it; to excel is to achieve a new height of performance and value.

Nature is excellent; Nature is the apex creator and life-giver; Nature is in most of its actions the pure and excellent life maintainer. It is never out or in a position of standing without any relation to others. Nature is in a constant connection to all that it manifests and to all that it has a relation to, with pure excellence.

My intention is to ignite my Factory of Manifestation with Nature's related words' for it to nurture sufficient outcomes visible within the spectrum of my physical environment. From now on I will keep the word outstanding where it belongs: out, and I will excel with the natural power of the term Excellent.

Comment Suggestion

Nature is complete at any given moment. All stages of the process of manifestation, regardless of the subject itself, are pure and profound at all times.

Each and every stage is unique and perfect within its own stage, manifesting the process of life, manifestation, uniqueness and value. The dynamics of life and the everlasting and ongoing movement and development are harboured within each and every step and stage of manifestation.

Whilst Nature manifests a flower, all stages are sufficient and brilliant in their own profound fashion. Once a baby is developing in the womb, all stages are of necessity, of importance and of value.

I experience unique things in my life; some of us are painters, poets or music composers and some sketch new buildings and cities. Whatever I do and whatever stage I am at, during the process of creating and materialising my ideas into physical recognition, those stages are of greatest importance, of greatest value and validity.

Acceptance is the key characteristic of Nature and the way it maintains its relationship towards All That Is.

Beyond any pure recognition, Nature looks carefully and is dazzled, hypnotised by the pure state of all that it is composed of.

All that is Nature is accepted as pure, as innocent and as wonderful. If there were to be a place to offer any guideline for an additional improvement, perhaps a supportive idea, Nature would contribute to it from its overall profound wisdom and Suggest this; its idea.

The recognition of momentary completion, at every step of the overall progression of any part of Nature, is the basis for the relationship between Nature and all that composes it.

Suggestions for further expansion, advancement and growth become the constructive force and support the transition towards an improved level and state greatly. Nature does not, ever, Comment on anything that it is; as this can neither originate from a place of acceptance nor from a point of admiration.

When anyone looks at what I do or establish, they perhaps use the term comment after reviewing the subject at hand. The word comment is related to the source of non-completion and to the necessity of improvement, and therefore, sends signals of disharmony and dissatisfaction to the person who has established it.

I shall observe the feeling that I get looking at previous events when someone has approached me and stated that he or she has a few comments, and I will grasp it and study it. I am likely to observe that this event has not been filled with true acceptance and recognition of my work.

On the contrary, when somebody mentions that he or she would like to suggest something, recognition, acceptance and harmony are all around. If I cannot recall particular moments as such, I close my eyes and feel those two words and feel the vibrations that my body expresses whilst setting direct focus upon each of them individually. Suggestion is aligned with the Nature of interconnectedness and with the Nature of collaboration, trust and support.

When anyone would like to suggest something, it confirms that a great recognition of the value, of what I have established so far, is there, and perhaps a new path towards further development can be used and may be welcome.

When my actions are received as valuable and constructive to a certain degree, I attract more valuable energy to keep on going with my projects and/or ideas. I am thrilled that the value is already visible and that it has the potential of growing better and stronger.

This makes me keep expressing and building. It allows anyone to harness the essential force in order to manifest and complete whatever he or she is doing with ease and efficiency.

My words ignite either disharmony, which leads towards inaction and has no relation to Nature whatsoever, or harmony, which leads towards excitement, thrill and completion with a strong and powerful mindset. This is not the case when the term comment I used.

Nature recognises all stages as valuable, and therefore, maintains the process of manifestation and advancement. Suggestion is by far the better term to favour, and it is in pure alignment with Nature and the way it talks to me through continuous manifestation, development and expansion.

My suggestions and great ideas support further growth for anything that receives my ideas and points of action.

Acceptance also allows others to follow my Suggestions.

Obey Follow

Traits of manifestation are all around me. There are unseen potential lines of progress, which are the paths towards my future selves. Nature and all of its parts Follow the offered lines with pleasantness and joy. The lines are eternal and lead towards all probabilities.

In each and every moment, every part of Nature can willingly choose and prefer a particular offered and potential line of progress and step into it by free will.

Nature follows an unseen potential progression and does not Obey anything and no one. Nature follows willingly and so do I as I look into my organs and breathe. All of my organs follow spontaneity and maintain my physical body alive and healthy by simply following the offered traits.

When anything is in a position of obedience, and of the duty of having to obey, it intensifies its distance to the part that is about to manifest and that currently dictates the required necessity; this places its enormous energy and intuitive expression to rest.

Obedience does not imply power or the desire to express. It manifests the feeling of being detached. When I follow, I can sense the characteristics of togetherness, connectedness and joint collaboration.

I follow my senses and never obey them. Nature and all of its manifestations are similar and all of its parts follow its guidance and do not ever obey out of obligation or pure force.

The water follows the stream down to the nearest lake or ocean and does this joyfully and in clear agreement and accord to participate and to become one with the stream.

When writing down the word obey and focusing upon it for a few moments, I reveal that my feelings are the exact opposite of acceptance.

Even if it is something that I do agree with doing; when attaching the sense of obedience to the equation, unpleasant vibrations are gathered and disharmonise the entire structure.

When any of my actions are disharmonious, I will never act to the best of my capacity, and this applies to anyone in relation to anything. When I must obey, disharmony appears. When I follow, interconnectedness and the sense of being a-part-of arises and immediate harmony supports an apex performance.

I am the result of actions following a supreme available guidance that I have followed since the very first moment of my existence. All that which is part of me forever Follows this guidance, and therefore, the remarkable result of existence is always apparent and present.

Use Utilise

The material world and all that is present here and now are for the greater good and for the expansion of life and development.

The building blocks, which Nature has in order to manifest a balanced and advanced world, are willingly taking part in the overall manifestation and in that manner are very pleased to become what Nature intends them to become.

They all are here for the purpose of becoming. Nature does its constant and continuous manifestation with the intention of creating for the greater good and makes something better from what it has been doing up to now.

The fact that all the minute particles have the true joy and will to be a part of what I experience, indicates the characteristics of that motion: Utilisation. Nature does not Use, it utilises.

When I look at the term use, I reach the conclusion and the feeling of doing something unkind, and I allocate creative power away from the part that I have been using for my purpose.

Interconnectedness, harmony and oneness do then not reside within that event, within that feeling of using something or someone. However, when I have the intention of utilising something, this unique intention harbours the characteristics of and the relation to advancement and improvement for all sides involved within itself.

As trainers train with the intention of utilising the talent of their students in order to elevate their skills to higher levels and, during that, improve their own coaching and instructing abilities, so does Nature utilise particles and all matter substance in order to bring about a living world of events and magnificent movements.

Utilisation is a joint collaboration in order to maintain advancement and growth of all parties, whereas the use of something in most cases solely benefits one side and is not aligned with the true nature of Nature.

In order to become a greater performer, regardless of my occupation or vocation, I must embrace the enormous power of benefitting all sides when doing any action. The word Utilise is the most accurate and constructive word in this case and implementing it within any context will support it better and will lead towards the expected, desired and potential performance.

Statements such as 'I felt so used' or 'He used me' do not contribute to the overall health and happiness of any person, for it is already so strongly placed within our collective mind and perception as a non-constructive event with a directly pointing pointy finger on the word use.

I am an integral part of Nature and as long as I act and think like Nature does, I experience more alignment and harmony within my actions and my life as a whole. I make my words as compatible as possible with the desired unseen; however, a yet visible stream of life, my events and experiences magically demonstrate my awareness as valuable and highly beneficial.

Serve Provide

I live in abundance. All around me is a clear manifestation and translation of abundance. The sky, the rain, the sun and the stars are all abundance; visible to me in the spectrum of light in the physical realm. Nature Provides me with all of that and much more eternally, in pure agreement. Nature is the provider and never the servant, and therefore, maintains the characteristics of the Prime Creator with validity and power.

I am the provider of my life. I participate in the delicate and constant manifestation of my individual life as well as in the life of all others. Exactly like Nature, I am the provider, and I do not perform at a level of Serving.

I decide consciously what I would like to experience both on conscious and unconscious levels and witness the plan and premeditated actions born within the core of me and not that of others. Nature establishes the patterns of growth, the path of expansion and the capacity for advancement, and it gets those from its core and not from others.

Once I serve, I operate according to the orders and plans of another; however, when I provide, I participate in the entire process as well as the planning and the decision-making. When I describe my actions and assign the word serve to them or its related words, the entire mental structure of this particular description allows energy and power to flow out of my hands.

I get onto a level where I do neither have a high capacity nor force for expression nor maintain results and efficiency. It relates what I am and do to servitude, which is a very unconstructive connotation.

Whenever I feel that I serve, I feel in a lower position and that I do not have a sense of participation or a sense of playfulness and collaboration. Nature manifests and participates with playfulness and with prominent collaboration. I am Nature, and I provide, and therefore, serve does neither have an appropriate meaning nor a true relation to what I really do.

Abundance is felt when I know that I provide. I feel the idea of plenty, and I have the willingness to share with others what is in my possession; this may be knowledge, physical materials, supportive actions or constructive and life-promoting ideas.

I, as each and every one of us, am a true creator and the authority of making decisions related to personal physical existence. I can easily provide, for I have participated and determined what is ahead and what should be achieved.

I can merely provide and not serve, for I am plenty and I am abundant. I solely utilise the term Provide within any of my fashion of communication so that my life is aligned with the ideas of natural plenty and abundance.

Save Accumulate

The collection of eternal possibilities, potential growth and expansion is Nature's characteristic and its way of expression. I am in the midst of an everlasting movement and abundance. I live within all of that and become it as well.

Growth and expansion can be achieved by gathering knowledge, results, wisdom and various key supporting methods. Gathering and Accumulating are part of expansion and fulfilment.

To accumulate is to know that all is available and that plenty of resources and the essential building blocks of my life are available at all times. Once I Save, I believe that I come or have already come to a point where all is below the essential level of supporting life or joy.

When I begin to save, I direct my mind towards the idea of not having what is desired. It manifests itself in my life based upon the clear focus of a situation of not having what is required.

Accumulate is associated with having plenty, and it is on the way to becoming ever more. Save is associated with an unpleasant situation and the upcoming of not-so-many bright days and times. All the inner interpretation of words leads towards thoughts and thought patterns that lead towards life experiences.

I am plenty and riches. My capacity to get the right ideas and to pave paths towards success, health and wealth can never be counted. They are all eternal. I am the master of ideas, and thus, can bring to mind more when desired.

I accumulate and grow stronger as Nature Nature does. Utilising the very same terms as Nature by preferring the term Accumulate, supports the current state and leads towards greater and better times with ease.

I think of times when I have favoured the word save and evaluate whether any of these occasions have been related to plenty and abundance or whether they have rather provided me with feelings of the exact opposite.

My feelings guide me towards the most aligned words and ideas with Nature. Those findings, once followed, enrich life and provide spontaneity and a sense of liberation and happiness.

As a conscious being, I have the right to prefer one thing over the other. Preferring Nature's way of talking is the utmost natural way of speaking and thinking. Once my speech is aligned with the behaviour and ways of Nature, it implies that I utilise by far better and more life-supporting words.

Forgive Accept

The main reason for achieving the most relevant and essential balance between all things is the fact that all has the very same basic substance, and therefore, all is equal in the overall picture of life.

Once I dwell consciously upon that fact and the reality that is composed of the very same components, thus shaping life in magnificent varieties, the feelings of interconnectedness and of oneness are available and present. In case all is one and originates from the very same source of expansion and of life, then all is equal, and all must be accepted as it is, at this very moment.

Accepting is allowing my existence to become aligned with that of others. I accept all actions and activities as a valid part of life and as an important momentary expression of existence. This leads me to accept my own existence to the fullest, knowing that whatever I do or wherever I am at in my life, physically, emotionally and professionally, I am in pure alignment with All That Is.

Nature accepts all as it is and grants valuable existence to all parts as prominent pieces of the whole. All is valuable, all is sacred, all is pure and all is innocent at this very moment of life.

I am also valuable, sacred, pure and innocent at this very moment of my individual life as well as my existence as a whole; and so is all that I grasp and perceive with my various senses.

To Forgive would be to assume that the other's actions have been non-valuable, non-sacred, impure and not innocent. To forgive is to acknowledge that the actions of the other person can solely be settled with my intention and at my choice as if one has the right to define what is to be settled and forgiven, and what is not.

The word forgive describes a state of superiority and inferiority between two individuals and two parts of the very same manifestation. This is not what Nature is. Nature and all of its parts are completely equal and convey similar building blocks and consciousness.

Acceptance is the natural way of harmonising all sides, allowing them all to be gathered in joy and oneness, benefitting from the eternal and life-giving energetic force by the Prime Creator. Harmony sets all the receiving channels back in place; aligning all parts of the whole in the absolute and most beneficial position.

Accept is the term that aligns me and my dear ones, and all that is my reflection back into one godly state with the source; thus, allowing the flow of abundance, expression, spontaneity and health to be present and active. Once one asks for forgiveness, I offer acceptance. True Acceptance of all, which is out there, is the finest path towards supreme achievement and fulfilment.

Investigate Study

The power of the whole is to allow all of its parts to play a prominent and a valuable role within the overall structure. Nature, with all of its parts, identifies the validity of all that is contained within its body of existence. All the parts are contributing to the performance and the successful coexistence of them all.

In order to advance and expand continuously, Nature benefits from the potential of all that it contains. It observes and is amazed by the potent potential; from the smallest part up to the biggest one all is valid, valuable and creative. Studying the parts happens with the sense of looking and being amazed by the contained potential for expansion.

Studying allows me to observe the apparent other with pure harmony. Both of us are related and connected, and we still have the ability to study one another and allow potential growth and expansion to take place and be present at all times.

When an Investigation takes place, the sense of oneness is no longer present; the pure connection of life between all parts is set aside and harmony is not at its best capacity.

Most investigations have the intention of getting to know something that is unclear and seemingly unknown, and in many cases with the sense of the feeling that the inspected and investigated party did something that is unacceptable.

I feel the word investigate once it is pointed at me or another. I feel the word study when it is pointed at me or another. In most cases, I grasp the most harmonious feelings whilst dwelling upon study. It generates vibrations of coexistence, of clarity and of benefit from and to one another.

The more harmonious vibrations there are within my physical body, the more stability and clarity my mind and entire existence will have. My words are the true stimulation of my feelings and the condition of my body and mind. Nature is in a state of harmony and observation, studying the potential of its components and life partners.

I am part of that Nature and acting as such gets me closer to the most efficient way of living, studying both with my words and with my actions, being like Nature and mimicking its behaviour as much as I allow myself to. Investigating benefits me no longer, whereas Studying reveals the purity and the potent knowledge within the subject of my interest.

Secure Safe

The creative force of life resides in all and permeates through all that I can find within any reality. My very own private existence and expression of life is the very same as everything that I can vividly observe whilst utilising any of the available senses.

The substance that comes into life is similar and originates from similar sources, and therefore, I am in a state of Safety at all times. Life is pure truth; existence and expression are truth. I am in a state of safety at any moment of existence.

When I think about the idea of Security, I dwell upon inequality that represents a gap between all things. To secure is to protect one thing from another or something, to acknowledge that I can be unsafe whilst being in the presence of something else, and therefore, must be secured heavily.

Nature and all visible events are the translation and the manifestation of my mind. I experience that which I have related myself to prior to the physical impact. I live within my manifestation and allow myself to define that which is acceptable and that which is not. Nature is aware of the fact that all which exists is the result of a thinking process materialised into physical form.

All is safe according to Nature; for all has eternal existence and life of eternity. Nothing should ever be secured; for it is derived from the idea of being in a threatening situation. The stream of life through all parts of Nature is saturated purely with safety and oneness.

All is safe as far as Nature knows, and this is the most accurate language for me to hear and mimic. Safety is an elusive necessity, for all is safe at all times. Life is the absolute sole truth, and I am in a state of existence, continuously. When I set my attention upon the idea of security and the idea of Safety each in its turn, I can easily determine which of the two provides me greatly with harmonious feelings.

The vibrational streams derived from words and thoughts are the lines by which my future life is being envisioned and painted at this very moment. Once they are established in the most natural way, they lead towards great fulfilment and harmony.

I am the Thinking Substance of all that I know and my choice to be as natural as I may be, supports great advancement and richness. Worthiness is the key word for accepting anything and anyone as an entity with the full right of expression and fulfilment.

Powerful Statements

- Each and every word counts, and therefore, I favour the most harmonious words that signal the element of freedom-of-manifestation towards my subjective mind.
- Nature Allows me to grow freely, to explore, to manifest, to become whatever I would like to become and do, without any supervision, and with complete freedom of action.
- My personal self-image, which is dynamically harboured within my subjective mind, evolves through those traits of life.
- My personal image evolves on an ongoing basis with each and every thought and idea that is transmitted to the subjective realm.
- All of my actions are performed much better and very close to their fullest capacities once I have the feeling of Worthiness and the sense of value within my mind.
- Nature addresses the meaning of the word Afford through its expression, and this is the most harmonious way of allowing myself to acknowledge that my being here is appreciated and valued.

- The level of Affording is the true alignment of my ideas and mind-structure with the fashion of Nature's expressions and talk.
- Anything that we do with the very same characteristics of Contribution gets delivered with far greater results and electromagnetic vibrations; aligned with Nature's behaviour and strength.
- Each word has a vibration, and it could be constructive and life-supportive to a certain degree or not.
- I can make a conscious decision to allocate more vitality and harmony by using apex words.
- My intention is to ignite my Factory of Manifestation with Nature's related words for it to nurture sufficient outcomes visible within the spectrum of my physical environment.
- Suggestion is aligned with the Nature of interconnectedness and with the Nature of collaboration, trust and support.
- Nature recognises all stages as valuable, and therefore, maintains the process of manifestation and advancement.
- When I must obey, disharmony appears. When I Follow, interconnectedness and the sense of being a-part-of arises and immediate harmony supports an apex performance.
- Utilisation is a joint collaboration in order to maintain advancement and growth of all parties, whereas the use of something in most cases solely benefits one side and is not aligned with the true nature of Nature.

- Abundance is felt when I know that I Provide.
- To Accumulate is to know that all is available and that plenty of resources and the essential building blocks of my life are available at all times.
- Once my speech is aligned with the behaviour and ways of Nature, it implies that I utilise by far better and more life-supporting words.
- Nature Accepts all as it is and grants valuable existence to all parts as prominent pieces of the whole.
- Once one asks for Forgiveness, I offer Acceptance.
- True Acceptance of all, which is out there, is the finest path towards supreme achievement and fulfilment.

Solve Settle

When things are not in harmony with Nature or not in the most natural state, it is potentially the right time to align the condition with the natural one. This can be done with clear flow and acknowledgement of the natural state of things.

Calmness and acknowledgement are all that are required for the manifestation of harmony. Nature does this best when two of its parts are not in the most harmonious position, so it simply manifests a Settlement. This settlement happens through the manifestation of a new state where all sides are comfortable to continue with their endeavours.

Nature establishes this by knowing that all is required and each and every step of the way is essential and valuable. Once I have the desire to Solve something, it implies that my actions contain the derived idea that the current state and situation are not optimal and do not set pure awareness to the absolute donation of that state to the bigger story of life.

All events are valuable. All events are contributing to the advancement of all sides and to a strong continuation of life with the flow of expansion. All is good. To solve means in most cases not to see the goodness in all events and to act from a state where non-acceptance is present.

To settle is first to see vividly that the event is valuable and contributing and that a new state can become a valid condition. Moving forward in comprehending the validity of all events, even if they seem to be non-constructive at times, is the natural and the most harmonious way to progress with ease, whilst benefitting all sides along the path.

My words are thought supporters, and when I think of settlement whilst my subconscious behaves in a natural way and aligns approval with the state so far, I will receive a stream of thoughts that is related to the most beneficial potential outcomes and to the ways of reaching the best outcomes.

My inner self works at its best when it accepts the contribution of all steps until now; thus, acknowledging the validity and the value gained by all the event participants, for when acceptance and admiration are in place, solutions come in the most natural and easy way to the place where they are utilised.

I lead my mind to better solutions when I express the word Settle and when I leave the term solve aside. Granting an absolute appreciation for all that has been envisioned and manifested so far is Nature's way' of talking and behaving, and that is the supreme and most favourable mechanism in order to maintain harmony within any system.

Responsibility Privilege

Life is a gift, and I should better implement that idea strongly and profoundly within my entire system of thoughts and awareness. I am here, and I have been given the right to express my godhood within a physical system of existence: my life.

Nature expresses itself on a continuous basis and constantly brings more beauty and variety to the physical realm of manifestation. All the actions that Nature is engaged in are grasped by it as prerogative actions and as it being Privileged of having the ability and the capacity to become a creator and manifest.

All is designed and thought of with pure privilege and the idea of Responsibility always resides outside of Nature's spectrum and range of manifestation.

The joyfulness and the excitement within the manifestation, as I can observe it as a fortunate owner of my senses, come with a profound and intense prerogative activity. Nature is pleased, happy and honoured to manifest, to bring new fulfilment to the world and to express its capacities to the fullest, honouring God's will to create and support supreme life. In order for me to be called successful and efficient, all of my actions must be done with complete agreement and excitement on my behalf.

When I feel responsible, I have the notion of doing something without full consent and acceptance on many of those occasions. When full acceptance and consent are not visible and/or not part of the activities, success and efficiency will never be at their best.

I perform according to my full capacity once true consent, free will, acceptance, excitement and supreme privilege saturate all of my actions. In order to succeed in all of life's endeavours, to realise in perfect harmony and superb beauty, the entire system must be involved.

The system of the Complete Will to act and to perform the various tasks and actions is the most important one for high performance and accomplishment. When all parts work together, the movement towards successful results is reached easily due to the fact that one force combined with additional parts is in absolute agreement and collaboration.

When I count the things that are saturated with the idea of responsibility, and when I feel their results and performance, I reveal the true nature of the word responsibility. Once I look carefully and with full attention to the actions performed with full privilege, and once I evaluate their results both for myself and for others, it shows the absolute might of the sense of prerogative and privilege.

I find the divine powers that reside in the actions done whilst being in a state of full acceptance and excitement and done in a privileged manner. Privilege has absolute collaboration and agreement with each and every action, whilst responsibility does not possess the absolute agreement, and thus, can never perform better than privilege.

Ideas become life events, resulting, by observation, in what has been stored within the inner system and the mind before. The words and terms dictate the fine structure of each belief and lead me towards full capacity, apex performance and fulfilment once the idea of Privilege is the leading power of the performed actions.

Attempt Act

Successful actions and high performance are achieved with strong faith, collaboration and trust in any specific action.

Whatever I do always shows greater and more admired results once I possess faith in completing the action through the collaboration of both the inner and the outer system by harnessing all the various capacities for the very same vision, and when I have the feeling of trust in order to feel free and calm whilst performing anything.

Once all of this is in place, I observe spectacular results and high performance, and I experience amazement by anyone who observes and looks at my actions. A complete synergy of all my true powers is revealed once harmony is established and trust is in the lead.

Not all Attempts always contain pure faith in the completion of the task,' does neither ever hold a strong collaboration of all the available parts nor show true trust in Nature and in me.

An attempt has, of course, the desire for a good accomplishment; however, does not possess the absolute focus upon high and absolute performance. I attempt and feel that it could lead to either successful or unsuccessful outcomes.

It does not sufficiently nurture the idea of successful outcomes, and therefore, does not support as much as it could do if I had a direct and profound focus solely upon the desired outcomes, successful ones.

Acting like Nature does, is with complete trust, full acceptance and absolute collaboration from all parts. Nature nurtures the single desired outcome and maintains focus along the way until this very remarkable outcome is achieved, visible, able to be experienced and to be observed.

I act with full capacity once I, exactly like Nature, allow myself to set a clear focus upon the desired outcomes and results. My mind is guided by my thoughts; my thoughts are the unseen partners of my language and words. The utilisation of the word act generates the preferred thoughts, maturing my desires into precise outcomes.

Feeding my thoughts with the word act means feeding my thoughts with the derived ideas of trust, collaboration and faith. That, which the thoughts are focused upon, leads to beliefs and thus to life events and experiences. Once I feel the word attempt, I can decipher for myself how much faith and trust it contains to and for me in comparison to act.

Dwelling upon the word act and observing the quantity of faith and trust within it conveys its true potential. Acting as surely as possible and maintaining trust and faith in all my actions result in successful outcomes. All preceding ideas become life.

Once I can grasp the grand power of the idea of acting and once I begin any action with this very idea as the leading force, fulfilment will surely be granted. By intending to be as efficient as possible and performing more successful actions along my lifetime, I move towards the most advantageous results. This can purely be by acting. With every single action, I Act as if it was the sole probable probability for me.

Donate Share

True riches can solely be achieved once the feeling and the knowledge of everlasting abundance is the dominating belief of the person at hand. The realisation that there are solely abundance and riches in life can be comprehended once the right focus is set upon this one valid truth.

Nature activates the Formless Substance at every moment and knows that this supreme substance, which has a direct relation to all thinking processes, physicalises form following premeditated thoughts. Nature expresses and lives by the principle of true abundance and knows that all is available, and once I receive anything, all is still available to and for others. Nature Shares from its vast experience, knowledge and wisdom and never Donates.

Donation comes with the idea of giving something to the other with the acknowledgement that the other requires it and potentially cannot get it by himself or herself. This leads to a greater necessity and requires additional donations. The cycle of donation attracts more of it and perpetuates a state where abundance is not comprehended as the truth of life and of Nature.

I live better when knowing that all is available, and once I share, I increase the quantity and life for all. There is abundance all around me, Nature is all around me and Nature in purity is the natural expression of abundance.

I live life on the physical plane with solid and remarkable abundance and riches. I am riches and all of my thoughts have the potential to provide me with any of my desires.

The path towards this realisation is through intentional thoughts about abundance, riches and everlasting Formless Substance that expects me to materialise it in accordance with my thoughts. A clear focus upon thoughts of abundance promotes abundance within the individual who generates and holds those thoughts.

To share is to firstly comprehend that there is more at any given moment and to secondly fathom that there is a constructive increase and expansion to all, including the entity that shares.

I am like Nature, and I share. I let go of the term donate and become aware of the fact that all can become rich, and all can live the true natural state of life; all manifests its individual desires with ease and flow by thinking unique intentional thoughts and by focusing upon Formless Substance. I become aware of this truth, and thus, support riches for all others.

When something is shared, it encompasses the feeling of abundance, collaboration, trust and love, whereas when something is donated, the focus is upon the inability of others to get things by themselves; therefore, the aspect of sharing increases and supports life events to a greater extent.

Thoughts of riches, shared whilst being in the expressive state of mind and of life-supporting growth, abundance, true potentiality and successful and efficient actions, construct astounding life events and experiences.

Share with the notion of plenty, abundance, riches and with the idea of the availability of the evermore materialisation of things, events and life. I am a Thinking Substance, and my life is the exact reflection of my thoughts.

Practice Execute

The actions that I think about prior to their actual performance are all within my mind to begin with, and thus, do paint strong implications upon my thoughts, future events and experiences in the physical world.

Each and every possible action is envisioned and manifested in a place, following its initial manifestation process, which is the thoughts' realm. All is envisioned and realised either in realities that I am aware of and set my focus upon or in realms where I have decided not to set a clear focus upon by not utilising my additional available inner senses.

Nature is aware of the fact that all gets manifested to a certain degree, and therefore, has decided to perform each and every action to its fullest with the leading idea of momentary perfection. All is momentarily perfect and constructive to a certain level, and therefore, should be acknowledged as such.

This vital acknowledgement allocates feelings and ideas of constructive and contributing performance, which refines all of Nature's actions and promotes further increase and advancement as intended. I envision anything within the thinking mind and allocate potential Thinking Substance to act upon those thoughts and imageries. When I think of Practice, I do not give my full capacity, and therefore, whatever I practice never leads towards real advancement, increase or expansion.

When all actions are Executed to the very best of the individual's capabilities and saturated with the pure intention of increasing and advancing, I am always in the state of advancement and increase, which are both fundamental intentions of Nature and of life. Each and every action counts exactly as each and every thought counts, and this manifests life accordingly. I act in the most natural way when I know that all is manifested, all is physicalised, and all contributes to my advancement and progress.

Nature executes and instils all potential power in each and every action and manifestation, be it big or not. All manifestation is important and significant to the overall stability and progression of life and of the entire manifestation.

Practice implies setting aside prominent power and intention and acting below the available potential. This promotes a state both within the mind and the body of acting on those levels and that in turn generates convenient connections and relations solely to those levels.

Once the mind or body is comfortable and becomes related to specific levels of performance, it favours those levels and sets them as the ones to choose and to act on accordingly.

To maintain high levels of performance and of execution, I had better always act as if it is the most important action and time, and allocate all the potential creative resources and utilise them at the moment of action.

Both the mind and the body become related to the levels of action, and they become familiarised with them in relation to the overall capability of the individual. This sets a proficient level of performance and advancement, and thus, an increase becomes present and active.

The term practice presents an immediate possible level of reduced pace and performance. Our thoughts of practice convey pictures of performance that are not at their best, and our words and language increase the focus upon those not-high levels.

Whilst setting the focus solely upon execute, the thinking mind conveys imagery and vibrations towards the supreme level of performance; increasing the excitement, the anticipation and the faith of the individual and performing at the highest possible levels.

All performances are the reflection of the images and ideas within the Thinking Realm and are supported by the spoken words. Allowing myself to be like Nature by Executing everything to the best of my potency, is the path towards increase, advancement, excitement and pure performance.

Tolerance Acceptance

Free will is a right that all living things possess whilst experiencing life on the physical plane. I establish what I feel, experience and think about, for I am a living, Thinking Substance, which is materialised within the physical environment.

All that I interpret and decipher as events, is the reflection of my inner environment and as such I see what I am. When things are not to my liking, I often have the tendency to set this acknowledgement aside and do not relate those unpleasing events to the fact that I promote them continuously into my life, following my thoughts and beliefs.

As a wise person, I accept all that I pull into my physical reality by myself as a Thinking Substance, and this private Acceptance sets me free; thus elevating my state into a highly conscious individual who acknowledges the ability to draw life by free will. Tolerance has the nuance of not accepting within it, not truly acknowledging others or the apparent events and not assigning their existence to my own thoughts and beliefs.

I am the source behind all that I can sense, and so Nature manifests within the spectrum of absolute acceptance. Each and every person would feel better if the other accepted him or her and his or her unique expression and not if the other did their best and succeeded in simply tolerating his or her individual life expression.

Love is acceptance, and it leads towards personal health and happiness, for it establishes a free and harmonious system for me, the one who possesses and lives with complete acceptance.

In the case that I tolerate, I do not have a free and harmonious system within me; and health and success are not my true and natural life components.

To accept others is to allow them to accept me. Therefore, by accepting myself in the absolute best manner allows others to accept me by setting a profound focus upon pure acceptance.

Favouring the word acceptance and allowing tolerance to be out of any of my expressions, unspoken and unseen, establishes strong and valid thought tendencies, following my words and focus.

This constructive tendency streams and empowers the idea of solid acceptance within my inner system so that wonder will be seen outside in the physical environment, which is thus the direct reflection of my focus, thoughts, beliefs and inner environment.

Any minute of focus on my part, which I solely set upon acceptance, establishes constructive vibrations and profound implementation of true Acceptance within my inner system. This is an exquisite way for me to be satisfied and to maintain high calmness as well as a relaxed mind and body.

All that I have within my scope of thoughts is desires, goals, ideas and my personal expression, for I allow everything else to be as it chooses to be, respecting free will to express and manifest freely and deliberately.

Task Role

The way Nature brings life is reflected by its magnificent manifestations and remarkable enjoyments and by the continuation of manifestation and advancement.

Growth, which is Nature's main characteristic, is shown to me. With this observation, I can simply conclude that Nature enjoys expressing and maintaining an amazing path of manifestation. Nature adopts the Role of the creator, all that creates whilst doing all of that with a pure will and profound excitement.

Nature does not interpret the various actions as Tasks or anything that it is obliged to do. Full engagement and full recognition of the actions and the act of doing them are present in all manifestation activities and are better accepted and described by the term role. When task is the overall description of an action, I unconsciously dwell upon the undesired obligation of doing the task, and therefore, do not give my full creative energy as it would be had role been the overall governing idea of this action.

When taking up a role, a feeling of collaboration and ownership is present in my mind and actions. Becoming part of the creative power of any structure feels like being the Prime Creator.

Any of my thoughts of ownership assure that essential functions are performed with full commitment and excitement; playing a role and taking a role is becoming an essential part of my life, and therefore, it resonates with being part of the overall manifestation and being the creator itself.

The relationships of my ideas and my actions dictate the level of commitment, involvement and performance. Once I feel the enthusiasm and the thrill with each and every one of my actions, I retrieve the creative force with full capacity and the results mirror this unseen fact.

I am the entity that can and does allow a variety of streams to be streamed according to its premeditated ideas of the action at hand. When taking a role is in place, I allow a full stream and full support of the subjective world to be streamed and to saturate my mind and objective life.

Once I think of task, of obligation and duty, I feel as if these actions that I am about to execute are not truly something that I have chosen to do, so a gap is present; thus not resulting in a strong stream from the Collective Mind and the Prime Intelligence to support my actions.

Allowing myself to take a prominent role in my actions results in apex performance. The sole utilisation of the term Role supports greater involvement, excitement and fulfilment with each and every action that I deal with and perform.

Choice Decision

As I live in a reality of probabilities, I apparently experience purely one reality line and recall it and tag it as my lifeline. The truth is that all possibilities, all probabilities and all the potential options that I have considered throughout my life have been materialised and actualised in various realms of actualisation.

All is manifested and each and every one of anyone's thoughts is, in fact, turned into an actual event. I do not focus on all the potential; however, I focus on what I have eventually experienced and call it life.

Nature makes Decisions all the time, and it is aware of the fact that, in truth, all is actualised within a realm of objectivity. Therefore, Nature acts on taking decisions whilst acknowledging the fact that all the other optional line paths and possibilities will be materialised as well.

Nature does not Choose with the notion of preferring one over the other and does not think that the others have come to a place of no further growth. Nature knows and acts with purity and profound vision of all that has ever been thought of, materialised and become an actuality.

Once I act like Nature and become aware of the fact that there is solely something to decide and not to choose from, I will set a better flow towards life, towards taking up the next step and towards deciding what will be tagged as my next visible and aware experience.

This leads me to the freedom of thought and freedom to live outside of the illusory framework of time, space and form. The true comprehension of my nature and reality is knowing that all has turned out as it has been within my thoughts and state of mind.

To choose is not actually the reality of things. In fact, I decide upon actions; however, allow myself to trust the fact that I choose and by doing this affirm that the other probable and potential actions will not become my reality.

It is, to a certain extent, like a safety mechanism, which allows me to live peacefully when I have non-constructive thoughts, which I would like to neither have as my reality nor to be actualised. However, now I am at the stage of knowing that lessons and new insights are available for my constant growth towards higher wisdom and towards a more advanced way of living.

All thoughts become reality; therefore, it is a good idea to maintain intentional thoughts of things and events that I truly would like to have, experience and see unfold in physical reality.

Taking true ownership of my thoughts is an essential step towards comprehension and towards grasping the fact that I bring about my individual life following my thoughts, beliefs and ideas. Utilising the term Decision as the real truth instead of choice gets me closer to becoming as natural and as powerful as I can get.

Favouring this within my communications, writings and thoughts as much as possible and with complete awareness is a path towards creating a solid structure of intentional thoughts and powerful and intentional manifestation..

Take Guide

Nature inspires me with remarkable manifestation to which it is related and of which it is a significant part. Observing Nature exposes the very true nature of its behaviour and, whilst following this behaviour and manner of acting, I learn more about the true nature of myself.

The way I grasp Nature and the way Nature communicates with me is pure Guidance. Nature guides me towards the so-called unseen truth and towards life's wisdom. When I am guided, a full acceptance of my potential to walk and complete the path is held and respected.

Guidance implies merely allowing the other party to participate and to gain new knowledge currently harboured by the guide. I am guided by Nature, and my potential and capacity to successfully walk the path of life are within Nature's thoughts and behaviour.

I am capable of and successful in completing any path and any reality; and this, Nature knows and reflects. Once I am Taken by something or another, the profound acknowledgement of my capabilities and powers is not recognised by me.

When taken by someone or something else, I get ideas of non-superiority and of inability to act and succeed on my own. Thinking of times when I was taken to a place or to a conclusion, and therefore, now truly identify my momentary feelings with those back then, reveals the feelings of non-superiority and of necessity.

The careful feeling of those vibrations from those who took me and imprinted their opinion upon my state increases the state of non-superiority and shows how important the conveyed feelings of others are to me. Studying those moments where I was taken, as opposed to times when I was guided by others with true clarity, reveals by far more constructive feelings and acknowledgement of power.

Guiding me towards truth and towards a natural state of being and becoming is one of Nature's roles.

I learn by observing and when I truly behave like Nature and adjust my tools of communication (language) and of manifestation (thoughts), I reshape my inner self and my personal inner image according to the natural state of being. I also guide others and allow myself to be guided.

I accept the true and remarkable power and potential of all people and Nature around me and allow Nature and all people around me to accept my greatness, my capacity, potential and absolute and eternal power.

Once my focus is upon guide, and once the inner subconscious comprehension behind the term guide is implemented with conscious awareness, a strong and prominent stream of constructive thoughts promotes the focus and the tendency of thoughts.

This establishes a greater strength of the thinking self and encourages images related to the ideas derived from the term Guide. My acknowledgement of potential, capacity, ability and success is present, and those ideas will eventually receive support from the objective world that will empower my subjective realm and thoughts even further.

Compete Manifest

Nature's greatest lesson for me is perhaps the following one: I am here to Manifest and not to Compete. There is no competition, purely manifestation. I am the part that manifests and the part that observes anything that might be interpreted as competition.

The absolute and unique truth is that I manifest, and that is merely within the realm of the illusion. There is solely manifestation and Nature manifests continuously. It allows the best powers to be focused upon the further development and value fulfilment of each and every aspect of Nature's offspring.

When I compete or believe that I am in the midst of a competition, I step aside from the absolute truth of reality; the truth of oneness. All is connected; all is related; and all is one. Once a seeming disharmony from the source is in place due to the idea of competition, I have no longer clear access to the full available capacity of manifestation, of life.

When dwelling upon parts outside of myself that may aim for the very same goal and may potentially get there first, I promote the necessity for hurry and for competition.

Therefore, I have the idea that I should better hurry and compete. Whilst being in this state of mind, the calmness, which is an absolute necessity for profound manifestation, is no longer present.

When I am no longer calm, the actions, which I perform, lead to a state of mind with a focus is upon responsibility and the urgency of doing something. I have then left aside calmness, one of the most essential factors for creating life with harmony and with awareness.

It is great to move fast; however, with calmness and with the knowledge and awareness that time and competition are misleading fruits of the mind, and therefore, better be set aside at all times.

Nature manifests with ease. Nature brings new manifestations to life with calmness and joyfulness for the reason that it knows that there is plenty, and plenty is for all and eternal. This is the difference between creating and competing.

When all is one, and there is plenty, and time is finally off the table, I utilise my finest capacities, and I manifest the absolute and exquisite manifestation to the apex of my abilities, whilst manifestation, and merely manifestation, is the leading force of my existence. I become the creator and live on the creative plane.

Once the idea of competition and of being competitive has been removed from my mind and my language, my life shows the supreme powers of Nature and of existence.

I am to be like Nature, a creator with complete awareness of the idea of oneness and of interconnectedness. My ideas of acting as the creator and purely through manifestation support my overall success in life and my full personal expression.

I am grateful to observe people like myself and with similar interests on various and similar occasions. A common interest in a specific area increases the passion, the desire to advance and the desire for further growth. This is a blessing for myself and my current interests.

Whatever I ask for myself, I should better ask for others in order to allow greatness and true value fulfilment. Being a sole creator and acting as a creator, who sets an essential focus upon Manifestation, provides the very best results. I always look forward like runners do, aware of my actions, my thoughts and my desires.

Origin Source

Nature has a spectacular attribute namely that of being in all places simultaneously. There is existence everywhere, and there always has been in terms of true and natural terms of life.

I accept the fact that time does exist according to my agreement with its existence, and I operate according to this perception. The fact becomes clearer once I look into the way Nature talks to me. Existence is, always has been and always will be. There is no Origin, merely a Source of life and matter.

Origin points towards a beginning and an ending, a start and a final line; yet, this does not apply to my personal life and to the existence of All That Is.

There is purely life and continuous existence, and therefore, it has a source and not an origin. In order for each individual to live in the natural way and be aligned with Nature and the true way of existence, I should better teach my conscious awareness to become free of the perception of time and space to a very extent degree.

I will then live life and not manage it according to the idea of the time between the moment of my birth and my final moment in my current physical body. Thus, living life becomes allowing true value fulfilment, broad space for expression, manifestation, meaning and harmonious actions.

Living with the idea of a source aligns me with the insight that all, which I observe and envision within my mind, has emerged simultaneously and with an eternal connection. I am all that I see and observe. I manifest all that I see and observe, and all has a collective source, a collective mind and a collective energy. I am one and the natural way of things is a simultaneous existence.

The illusive ideas of both space and time are replaced when the focus is set upon the source of life and the collective life of all. Once I read the word origin and the word source and then observe the vibrations of the two, I witness the more harmonious vibrations that the word source provides me with; a profound relation to the eternal life is established.

I focus upon source and feel my current body conditions whilst setting a profound intention upon the collective source of life.

Once the source is in place and I choose the term source with the related ideas of a collective existence and of true freedom of life where time is all now and eternal, my expression of uniqueness begins to grow further and greater; thus, showing the exquisite capacity of expressions and ideas as if the layers of time and space have been completely removed, allowing freedom of expression and of manifestation become active in their full power.

For a couple of moments, I imagine the possibility of eternal existence where all is in a state of life and expression. I allow the moment to provide the meaning of this state of mind and the relation to the source and the collective mind, which relates me with all. Nature talks to me through the idea of a Source and from an everlasting existence and life; so, I may do that as well.

Demand Require

All is available; all the resources of life are truly waiting for the mind to present its new ideas and new visions at every moment. Once a vivid idea is in place, the everlasting and available energy for manifestation takes its role and aligns itself with the imagined visions of the mind, entering with grace into the realms of physical dimensions and presenting the materialised new part of life.

As long as I set my mind upon a profound focus and concentration, I am supported by the eternal available Conscious Units, which materialise on a continuous basis, leaping from invisible existence into perceived ones.

I am surrounded by abundance, harmony and the strong will to realise more and more by the Conscious Units, which can be referred to as the foundation of all life that I observe by constantly utilising my senses.

Nature is truly aware of the fact that the building material is available and eternally created and that once a mental vision is in place, the Required parts to complete this manifestation are always available. The truth is that the idea once harboured within the creator's mind, is already creating all that is most suitable and required for the completion of the manifestation.

Nature does not Demand anything from the source, it knows that the material is already there and is therefore available once the mind of Nature has been set upon it and agreed upon a new event, a new path.

When I demand, I operate on a competitive plane and in a competitive manner, for I believe that the available resources are not available on an eternal basis, and therefore, I demand and ask in a competitive fashion to get what is essential for the completion of any idea and vision.

Once my mind is set upon the creative plane and manner, I truly comprehend the idea of everlasting and ever-available material for the materialisation of my visions, and I become attuned to the true and natural source of things, which continuously provides immediate resources for all of my ideas and thoughts.

I do not take anything from anyone, and once I live by the creative mode, the alignment with true abundance presents itself within all of my manifestations and visions.

I require, exactly like Nature does, for I know that it is available. I do not demand, for that would indicate that the source cannot continuously provide me with all that I can and would like to utilise. I am the result of my thoughts and beliefs, and therefore, I navigate my life on this physical plane to an absolute alignment with the natural things of Nature, whilst requiring things and resources with the notion and the true comprehension of everlasting availability of all that I Require.

Work Operate

Action is the most essential activity in order to maintain growth and advancement, which life requires on the plane of existence, to continue.

The activity must be present during the life and during the manifestation of new visions and new events, and therefore, all participants should envision whilst acting.

Often I tend to relate action to Work and, even more, perceive my life activities as work and not as the sum of various contributing activities. When the word work is present, I usually set aside the attribute of freedom and keep those two concepts away from each other.

Work is an action in order to survive and to be able to maintain physical existence, whilst Operate is an act of freedom; the part of life where I can truly enjoy and express my visions and savour the pleasantness of life and all of its varieties.

Not all people are at this stage. Some are as pure and natural as Nature and have the perception that life is life and all activities are valid and contribute to one's progress and advancement and that the natural way of doing everything is the main reward and the reason for existence. Nature operates and acts in order to express.

The way all of those actions are perceived, and are thus by far more efficient, is a continuous action, constant growth and activity where all and everything brilliantly contribute to the individual part of Nature and to Nature as a whole.

When Nature operates, it accepts all of the minor and bigger actions and activities as equal, for each and every leaf is of utmost importance to the tree and to the entire forest. All is of high importance and significance.

When I act through operation and when I act the same as Nature, I assign greater value to each and every act of mine, reaching a point of total acceptance and living among the high value of all of my life pieces.

This grants me the absolute best ground for living in true freedom and in total harmony with everything and everyone around me. Once I live in profound harmony and all the actions, which I take part in, are of high value, I truly live in a godly manner, comprehending and existing through the eyes of the creator.

All, that is around me, is the pure result of true acceptance and profound harmonious actions, which are the result of the attribute of operation. Operation communicates high value, the agreement of action and quality.

Each and every word plays a significant role within the manifestation of my mind and its character, and therefore, once the focus is purely upon Operate, the essential energy is already implemented within the activities, the overall performance and success and is derived from those factors and the implemented source of manifestation.

Different Unique

All things are to a certain degree one of a kind, and often, for the purpose of easiness, I grasp them as similar whatever they may be. Whilst observing Nature and looking at a group of wild lions, I tend to believe that they are all similar to a certain extent, and the general picture allows my mind to grasp the idea of a wild lion in a much easier way.

As I look deeper into the true nature of things, no snowflake is identical to the other; no face is identical to another, even in the case of identical twins. The term identical really solely exists for the sake of simplicity of comprehension.

All things, all objects, all events, all items and all thoughts, which I envision within my mind, are not like the other. They are rare and Unique.

When describing something as Different from the other, I match the two, establish an unseen competitive comparison and evaluate one as better than the other.

Each and every moment of my life is not the same as another and has no capacity to be the exact replica of another moment; however, all contributes to the overall idea of what I describe and grasp as my life, and therefore, all is valuable.

Once a competitive comparison is in place, the absolute value of something is no longer; and for the observer and the one who conducts the competitive comparison, it is not present anymore.

For success and for maintaining the value of all things and for the absolute best experience and contribution to anything that exists, I had better identify things as rare and as unique within their kind. This approach maintains the true value of what the observer has in mind and streams the best energy and value towards the life experience of the observer.

I elevate my life value simply by acknowledging the value of things whilst assigning all their rareness and uniqueness to them. The expressive mind is the one that knows the constant manifestation of things in a various and unique single fashion. Whatever has been realised so far, will always be manifested in an absolute way. Life's evolvement is the key to ever-existing new manifestation; thus, new energy is harboured within all that I sense.

All the events that I have had and the upcoming ones that I will experience are aligned with the true nature of life and are all unique and constructive for my indefinite development. Talking and thinking like Nature restructures my mind to the natural way of living, observing and experiencing.

I am the master of my mind and the artist of my life and can and do paint it in the most valuable way when I acknowledge the true uniqueness of things. Allowing myself to describe something as Unique sets the foundations for the creative mind, which leads towards a life filled with absolute value and zest.

Goal Vision

Nature is the ultimate provider and the sole creator of all that I see around me. What I observe is the manifestation of Nature and the result of conscious and eternal matter materialising into physical form at a specific moment and place.

All that I witness around me is the result and the reflection of what I believe to know about the world around me.

I observe Nature through my physical senses, and I see what I am. Nature around me behaves according to my thoughts, beliefs and inner structures that I have established along my lifetime. What I see is what I am, and therefore, the way I observe Nature is the way I reveal myself to myself.

I constantly look at the living and breathing three-dimensional mirror and believe that what I see is not me for the reason that my physical body, which I relate to and refer to as me, takes part in that big miraculous production. Since I am Nature and Nature behaves in a specific manner and follows precise patterns, I am also wired in the very same fashion. Expansion and constant development are not part of my nature; however, they are what I truly am as a whole, Nature.

One of Nature's finest capabilities is to establish and fashion great and cohesive structures. It would not allow into existence a single tree without seeing the entire forest. Every leaf is part of the greater Vision, the forest.

There are no Goals, for goals are merely representatives of minute and partial pieces of a vision. Nature is a visionary and the way it moves forward is by realising visions with great flow and wholistic comprehension. There are no parts, there are solely wholes. There are no leaves, purely forests.

Vision is the way of manifestation, expansion, movement and growth. Once I think of a goal, I set my focus upon a single part of a grand and major picture. It is as if a painter sets a profound focus upon a single item within the painting without seeing the entire vision.

A vision is a way of grasping the intimate connection and life between all parts; allowing them all pure collaboration and manifestation; setting them in a position where they support each other and are supported; allowing them to be utilised and to utilise the rest of the items around them. I establish vision and the way I see in a clearer way is by placing my focus on the entire grand and major picture, the grand vision where all parts complement and support one another.

Goal is a conscious perception derived from the competitive aspect of life that does not advocate the creative mode, for its foundations are solely based upon the survival instinct. Nature does not have survival instincts; it possesses creative powers and comprehends the most valuable insight that all is eternal, all is for good and all is alive and stays that way.

The sole way of navigating through life and the various realities is by operating through the expressive mode, where the personal expression leads every action and where vision has its validity and capacity to realise with ease, safety and assurance that whatever is being manifested is constructive, valuable and meaningful. I am Nature; I am the utmost expressive force; I have the capability to manifest through Vision; thus allowing the greater picture to take place within my physical realm of expression.

Achieve Materialise

I am in a state where at this very moment I have a new vision in my mind, a picture that I would be very happy to experience as part of my physical life. All is available, and I am in the midst of manifesting. Once I utilise the idea of Achieving any new thing and have it as part of my life, I tell my subconscious mind within this statement that I am about to get something, and this would require a sort of effort on my part.

The idea of achieving anything is an idea that places me at a specific point and the vision of a distance, far away from where I am right now. All I better do is to find the right path towards it and whilst I am there, I am an achiever.

The way Nature realises anything is by manifesting from the here and now. The Here-element of manifestation by Nature indicates the centric position of Nature where all manifestation is taking place. Nature manifests from the Here. Nature manifests from the *now*. There is no *there* when it comes to anything that has been manifested by Nature, and there is no later when it comes to any tiny object that will be established by Nature.

Nature Materialises all the physical manifestations that I can sense as a valid part of this reality. I sense with my sharp sensors the elements of Nature that have been materialised within a centric point. The materialisation of things illustrates the movement of life.

The picture shows the creator within a solid and integral centre and all that is created emerges from this centre outwards into all directions. The manifestation is from this middle core of existence and life. I am like Nature. The moments in my life are all the result of manifestation from my centric point of life. They all emerge from the centre of my existence outwards into all directions.

I materialise all that I experience. I do not achieve anything, for all comes straight from my core of being, my middle space within existence and life. I materialise my physical body, and all of my so-called future events are all taking place at the very same place and time where I experience my present moment. *All is here. All is now.* All is materialise from a centric point.

I cannot achieve anything, for I have no place to go to in order to get it. Once the time is right for anything to be part of my life, it is materialised within the *now* and within the *here*. Here is where I manifest, *here* is where my life is materialise. I am the force of materialisation that operates at this very moment and at this very place.

I prefer to comprehend the manner of Nature, for I am one with Nature and as a part of it I do possess similar characteristics as Nature does, and one of the most important of them is the fashion by which my life is being manifested: Materialisation.

Powerful Statements

- To Settle is first to see vividly that the event is valuable and contributing and that a new state can become a valid condition.
- All is designed and thought of with pure Privilege and the idea of responsibility always resides outside of nature's spectrum and range of manifestation.
- I perform according to my full capacity once true consent, free will, acceptance, excitement and supreme Privilege saturate all of my actions.
- A complete synergy of all my true powers is revealed once harmony is established and trust is in the lead.
- My mind is guided by my thoughts; my thoughts are the unseen partners of my language and words.
- Nature activates the Formless Substance at every moment and knows that this supreme substance, which has a direct relation to all thinking processes, physicalises form following premeditated thoughts.
- There is abundance all around me, Nature is all around me and Nature in purity is the natural expression of abundance.
- All is momentarily perfect and constructive to a certain level, and therefore, should be acknowledged as such.

- To maintain high levels of performance and of Execution, I had better always act as if it is the most important action and time, and allocate all the potential creative resources and utilise them at the moment of action.
- Allowing myself to be like Nature by Executing everything to the best of my potency, is the path towards increase, advancement, excitement and pure performance.
- Any minute of focus on my part, which I solely set upon Acceptance, establishes constructive vibrations and profound implementation of true Acceptance within my inner system.
- When taking up a Role a feeling of collaboration and ownership is present in my mind and actions.
- Becoming part of the creative power of any structure feels like being the Prime Creator.
- All is manifested and each and every one of anyone's thoughts is, in fact, turned into an actual event.
- The true comprehension of my nature and reality is knowing that all has turned out as it has been within my thoughts and state of mind.
- When I am Guided, a full Acceptance of my potential to walk and complete the path is held and respected.
- Guiding me towards truth and towards a natural state of being and becoming is one of Nature's roles.
- I am here to Manifest and not to Compete. There is no competition, purely manifestation. I am the part

that Manifests and the part that observes anything that might be interpreted as competition.

- Nature Manifests with ease. Nature brings new Manifestations to life with calmness and joyfulness for the reason that it knows that there is plenty, and plenty is for all and eternal.
- I become the creator and live on the creative plane.
- I am one with the absolute Manifestation mode of Nature and of all existence.
- Nature talks to me through the idea of a Source and from an everlasting existence and life; so, I may do that as well.
- I am surrounded by abundance, harmony and the strong will to realise more and more by the Conscious Units, which can be referred to as the foundation of all life that I observe by constantly utilising my senses.
- I am a Thinking Substance, and my life is the exact reflection of my thoughts.
- The truth is that the idea once harboured within the creator's mind, is already creating all that is most suitable and required for the completion of the manifestation.
- Nature Operates and acts in order to express.
- When Nature Operates, it accepts all of the minor and bigger actions and activities as equal, for each and every leaf is of utmost importance to the tree and to the entire forest.
- All things, all objects, all events, all items and all thoughts, which I envision within my mind, are not like the other. They are rare and Unique.

- The expressive mind is the one that knows the constant manifestation of things in a various and Unique single fashion.
- Talking and thinking like Nature restructures my mind to the natural way of living, observing and experiencing.
- Allowing myself to describe something as Unique sets the foundations for the creative mind, which leads towards a life filled with absolute value and zest.
- Nature does not have survival instincts; it possesses creative powers and comprehends the most valuable insight that all is eternal, all is for good and all is alive and stays that way.

OBSERVATION
ALIGNMENT
EXPRESSION

Danny Oceans